Tarquin Sudoku

Logical puzzles to test your reasoning powers and how to create them

Gerald Jenkins

Contents

Tarquin Publications

Where did Sudoku come from?

These puzzles first became very popular in Japan where the structure of the language is ideographic and does not permit the kind of verbal crossword puzzle which are so well known in countries which use languages of European origin. Sudoku puzzles have since spread to all parts of the world and millions of people now enjoy a daily 'fix'. It is commonplace in Japan, as elsewhere, to see commuters on their way to and from work deep in thought as they wrestle with the logical thinking necessary to solve them.

Rather curiously these puzzles were not invented in Japan but probably by Howard Garns in the United States in about 1979. In the USA they are still known as 'Number Place Puzzles'. However, the idea and the concept is an old one and is as we shall see, a direct descendent of the Latin Squares, first described in classical times and since used for agricultural trials.

Latin Squares

Here are examples of 3 x 3 and 4 x 4 Latin Squares. Notice that each row and column contains just one of each of the letters.

Such patterns are used to test the yields or disease-resisting qualities of agricultural crops. By spreading the plots of each variety across the field in this way and averaging out the results it is possible to reduce the effects of micro-climate or soil differences. In a situation where there is so much random variation, the use of Latin Squares helps with assessing which is the best yielding or most pest resistant variety.

a	b	c
c	a	b
b	c	a

a	b	c	d
d	a	b	c
c	d	a	b
b	c	d	a

An extra property

If the second and third columns are interchanged on the 4 x 4 Latin Square, something very interesting happens. Not only does it retain the row and column properties of the earlier version, but each smaller 2 x 2 block also contains one each of the four letters. This additional property is only possible when the sides of the square are themselves perfect squares, i.e. when they are 4 x 4, 9 x 9, 16 x 16 and so on. It is immediately evident that this type of pattern would be even better than the ones above for agricultural trials because the plots are more evenly spread out across the field.

If we now substitute the values a = 1, b = 2, c = 3, d = 4, into the 2 x 2 version, we have what looks like a kind of mini Sudoku Puzzle. In each row, column and 2 x 2 block there is one each of each of the four digits.

Since it is possible to substitute the four numbers in different orders and so obtain 24 different 4 x 4 squares, we can create a variety of different 4 x 4 Sudoku puzzles by blanking out different selections of the numbers in them.

a	c	b	d
d	b	a	c
c	a	d	b
b	d	c	a

1	3	2	4
4	2	1	3
3	1	4	2
2	4	3	1

Two puzzles to solve

Here are two examples of such puzzles. For each of them, just four squares have been filled in. The puzzle is to try to fill in the rest using just the simple rule that each row, column and block of four must contain one each of the digits 1, 2, 3, 4. It is recommended that you try to solve both of these before continuing.

	3		
			2
		4	
4			

1		3	
	2		4

To a 9 x 9 Sudoku Solution Square

No doubt you found those two 4 x 4 puzzles rather too easy and too predictable to be of particular interest. However, such puzzles are very easy to make up and might well be of value in certain teaching situations. Also they could be used perhaps for a very young child who wants to do a Sudoku puzzle along with older siblings. However, most people will undoubtedly be looking for something which is rather more demanding and this leads us to the next one which has that extra square block property, the 9 x 9. It will immediately be recognised as the one which is used for classic Sudoku puzzles.

a	b	c	d	e	f	g	h	i
d	e	f	g	h	i	a	b	c
g	h	i	a	b	c	d	e	f
b	c	a	e	f	d	h	i	g
e	f	d	h	i	g	b	c	a
h	i	g	b	c	a	e	f	d
c	a	b	f	d	e	i	g	h
f	d	e	i	g	h	c	a	b
i	g	h	c	a	b	f	d	e

1	2	3	4	5	6	7	8	9
4	5	6	7	8	9	1	2	3
7	8	9	1	2	3	4	5	6
2	3	1	5	6	4	8	9	7
5	6	4	8	9	7	2	3	1
8	9	7	2	3	1	5	6	4
3	1	2	6	4	5	9	7	8
6	4	5	9	7	8	3	1	2
9	7	8	3	1	2	6	4	5

On the left is a Latin Square with one each of the letters a to i in each row, each column and each 3 x 3 block. It can easily be observed that the letters are arranged cyclically in blocks. On the right is the same square converted into numerical form by substituting a = 1, b = 2, c = 3, etc. The result is a 9 x 9 Sudoku square which has the special property that the first nine digits are ranged along the top row in numerical order.

Labels and Families

At first sight this new Sudoku Solution Square looks rather different from the one above but in fact it was obtained by substituting a = 7, d = 5, e = 1, g = 9 and i = 4 into the same Latin Square. The values for b, c, f and h were left as before. This emphasises that the numbers in a puzzle are simply 'labels' attached to the letters in the original Latin Square.

All the Solution Squares which can be derived from a single Latin Square can be thought of as a 'Family' and it is easy to calculate how many members it has.

For instance a could have been each of 1, 2, 3, 4, 5, up to 9, giving nine different labels for a. The letter b could have been any of the remaining 8, c any of the remaining 7 and so on.

7	2	3	5	1	6	9	8	4
5	1	6	9	8	4	7	2	3
9	8	4	7	2	3	5	1	6
2	3	7	1	6	5	8	4	9
1	6	5	8	4	9	2	3	7
8	4	9	2	3	7	1	6	5
3	7	2	6	5	1	4	9	8
6	5	1	4	9	8	3	7	2
4	9	8	3	7	2	6	5	1

The size of this family would therefore be:
9 x 8 x 7 x 6 x 5 x 4 x 3 x 2 x 1 = 362,880

It is more normal to write this as 9! or factorial 9.

Transformations

Any one of these 362,880 different Sudoku Squares can be transformed into further ones by interchanging row and column blocks, the rows within any row block and the columns within any column block.

None of the permutations ABC, ACB, BAC, BCA, CAB, CBA or 123, 132, 231, 213, 312, 321 (for each column or row block) destroys the properties of the Sudoku Solution Square. After any of these transformations, each row, each column and each 3 x 3 block will still contain one each of the digits 1 to 9. It is well worth noting that any number of these transformations can be applied in any order.

Be careful not to exchange single rows or columns with those in different row or column blocks.

How many different Sudoku Solution Squares are there?

$$9! \times 72^2 \times 2^7 \times 27{,}704{,}267{,}971 = 6{,}670{,}903{,}752{,}021{,}072{,}936{,}960$$

At the time of writing there appears to be a consensus that the number above, first output by a computer program written by Bertram Feigenhauer, is the correct number of solutions. It is a simply enormous number, enough incidentally to provide every one of the six billion people on the planet with a distinct Sudoku Puzzle every day for about 3 billion years! It is no wonder that newspapers have no problems in providing at least one new one every day!

In fact this number is so large that considerable ingenuity is needed even to determine the number of solutions with the help of a computer. The argument to arrive at this result is by no means easy to understand but we can at least get the flavour of it. We know that the 9! comes from relabelling the letters in the underlying Latin Square. One of those re-labelings can make for example the top left block always the same. It follows from this observation is that it is only necessary to count the number of Sudoku solutions which always have this one same block. Then, if the top left block is fixed it is still possible to swap the remaining six rows and columns in the ways described above. There are 72 ways (3! x 3! x 2!) of rearranging the rows and similarly 72 ways for the columns. This is where the second factor of 72^2 comes from. By exploiting the properties of relabelling and row and column swapping the number of solutions which need to be found by computer has been reduced by a factor of more than a billion. However, further ingenuity is still needed to reduce the search further to the point where even a powerful computer can tackle it. The final two factors, one of which a a huge prime, are the result of the computer search.

The Internet offers lots of sites and discussion forums which can be accessed to look for the latest developments in this field.

How many 'Givens'?

To make a good puzzle from any one of these Solution Squares, the puzzle setter who wants the puzzle solver to have an enjoyable and interesting time, has to choose a certain number of givens. If there are too many, then discovering the missing numbers becomes too routine and boring, rather like solving the 4 x 4 puzzles overleaf. However, if there are too few, then the solution can become very difficult indeed and may not be unique. There could even be millions of possible solutions.

For normal Sudoku Puzzles the balance appears to be struck within the range of 27 to 31.

Creation by Deletion

Having discovered that an ideal Sudoku Puzzle has to have between 27 & 31 givens, it might be thought that the best way to create one is to take a Solution Square and then to delete just over 50 numbers at random.

In fact, Puzzle J was indeed created in just this way. Exactly 50 numbers were deleted, leaving 31 givens. It turned out in practice to be quite easy to solve. Puzzle K has only 30 givens, as a 1 in the square marked with grey was deleted from Puzzle J. It is a little harder to solve.

However, when Puzzle L with 29 givens was created by deleting a further 1, also marked in grey, it no longer proved to have a unique solution. It has four different ones and is useless as a Sudoku Puzzle.

Unfortunately it is not possible to know in advance how difficult a puzzle created by this method will be or indeed that it will definitely have a unique solution.

Solution Square

5	4	1	8	9	7	2	3	6
2	8	9	6	1	3	7	5	4
6	7	3	4	5	2	1	9	8
4	9	5	2	3	6	8	7	1
1	3	8	7	4	9	6	2	5
7	6	2	5	8	1	9	4	3
3	1	4	9	2	8	5	6	7
9	5	7	1	6	4	3	8	2
8	2	6	3	7	5	4	1	9

J

							3	
2		9			3		5	4
		3		5	2		9	8
								1
			7		9		2	5
	6	2				1		4
3	1			2	8		6	
	5	7	1	6				2
				5				

K

							3	
2		9			3		5	4
		3		5	2		9	8
								1
			7		9		2	5
	6	2				▦		4
3	1			2	8		6	
	5	7	1	6				2
				5				

L

							3	
2		9			3		5	4
		3		5	2		9	8
								1
			7		9		2	5
	6	2				▦		4
3	1			2	8		6	
	5	7	▦	6				2
				5				

Creation by Transformation

Fortunately there is a much more reliable way of creating good Sudoku Puzzles of the required level of difficulty, The method is to transform one that you already know is of the right level of difficulty and has a unique solution using a selection of the transformation methods shown opposite. However much the Puzzle is transformed, it still remains essentially the same.

Since we knew that Puzzle J was a good one, we used it to create others. Puzzle M was created from Puzzle J by interchanging the first two column blocks and also the first and third columns in the third column block. Then Puzzle N was created from Puzzle M by increasing the even numbers by 2, remembering that 8's become 2's. Finally Puzzle O was created from N by increasing all the numbers by 1, not forgetting that 9's become 1's. The first two row blocks were then interchanged.

J

							3	
2		9			3		5	4
		3		5	2		9	8
								1
			7		9		2	5
	6	2				1		4
3	1			2	8		6	
	5	7	1	6				2
				5				

M

							3	
		3	2		9	4	5	
	5	2			3	8	9	
						1		
7		9				5	2	
				6	2	4		1
	2	8	3	1			6	
1	6			5	7	2		
	5							

N

							3	
		3	4		9	6	5	
	5	4			3	2	9	
						1		
7		9				5	4	
				8	4	6		1
	4	2	3	1			8	
1	8			5	7	4		
	5							

O

						2		
8		1				6	5	
				9	5	7		2
							4	
		4	5		1	7	6	
	6	5			4	3	1	
	5	3	4	2			9	
2	9			6	8	5		
	6							

This is a very powerful method and each of the groups of four puzzles A to H that follow were created in the same way. Teachers and puzzle setters can find the right level for a class or group by experimentation with the puzzles provided and then use this method to create new puzzles. The connection is unlikely to be spotted and even if it is, the puzzles are not spoilt or the solutions easier to find.

The Special Puzzles

To check your solutions to the special puzzles on page 39 it is necessary first to practice some simple transformations.

Simply interchange rows 1 & 2 and then column blocks A & B. Then compare with the solutions given here.

X

5	1	4	6	2	7	3	8	9
2	7	9	3	8	4	5	1	6
3	8	6	5	1	9	7	4	2
7	2	8	1	6	5	4	9	3
4	6	5	9	3	8	1	2	7
1	9	3	4	7	2	6	5	8
9	4	7	8	5	3	2	6	1
6	5	2	7	9	1	8	3	4
8	3	1	2	4	6	9	7	5

Y

4	1	5	6	2	7	3	8	9
2	7	9	3	8	4	5	1	6
3	8	6	5	1	9	7	4	2
7	2	8	1	6	5	4	9	3
5	6	4	9	3	8	1	2	7
1	9	3	4	7	2	6	5	8
9	4	7	8	5	3	2	6	1
6	5	2	7	9	1	8	3	4
8	3	1	2	4	6	9	7	5

Z

8	3	9	4	6	7	5	2	1
4	2	6	1	5	8	7	3	9
1	5	7	9	2	3	4	6	8
9	4	1	7	3	5	6	8	2
7	8	3	2	1	6	9	4	5
2	6	5	8	4	9	1	7	3
6	1	8	3	9	4	2	5	7
5	7	2	6	8	1	3	9	4
3	9	4	5	7	2	8	1	6

Tarquin
Sudoku

To solve this puzzle, you have to complete the grid below so that every row, every column and every 3 x 3 block contains one each of the nine digits 1 to 9.

Always remember that this is a completely logical puzzle and you never have to guess. Only write a number in a square when you are certain by logical thinking that it is the only number that will go there.

The working boxes are there to help you to organise your thoughts. You only need to use them to write down the missing numbers if you are stuck.

			5		4		9	
	4			7			6	
			8			5	1	
7	8			6				
	9			1			3	
	1		2		5			
8					3	5	4	
	7		1		2		8	
	5	6	7					2

There are 51 numbers to find.

6

Tarquin Sudoku

To solve this puzzle, you have to complete the grid below so that every row, every column and every 3 x 3 block contains one each of the nine digits 1 to 9.

Always remember that this is a completely logical puzzle and you never have to guess. Only write a number in a square when you are certain by logical thinking that it is the only number that will go there.

The working boxes are there to help you to organise your thoughts. You only need to use them to write down the missing numbers if you are stuck.

1						6		5
7				5			8	
6	2						9	
			8	9			7	
4				1			2	
				2		3		6
5		6	9					4
9				8		2		3
	3			6	7	8		

There are 51 numbers to find.

Tarquin Sudoku

To solve this puzzle, you have to complete the grid below so that every row, every column and every 3 x 3 block contains one each of the nine digits 1 to 9.

Always remember that this is a completely logical puzzle and you never have to guess. Only write a number in a square when you are certain by logical thinking that it is the only number that will go there.

The working boxes are there to help you to organise your thoughts. You only need to use them to write down the missing numbers if you are stuck.

6		7	1					5
	4			7	8	9		
1				9		3		4
2						7		6
7	3						1	
8				6			9	
		9	1				8	
5				2			3	
				3		4		7

There are 51 numbers to find.

Tarquin
Sudoku

To solve this puzzle, you have to complete the grid below so that every row, every column and every 3 x 3 block contains one each of the nine digits 1 to 9.

Always remember that this is a completely logical puzzle and you never have to guess. Only write a number in a square when you are certain by logical thinking that it is the only number that will go there.

The working boxes are there to help you to organise your thoughts. You only need to use them to write down the missing numbers if you are stuck.

	2		7		8		6	
9		8		5				1
		1	2				5	4
			3				7	8
			8	4		2		
		7	9			1		
	1	2				9		
		3	6			4		
		4					8	5

There are 51 numbers to find.

Tarquin Sudoku

To solve this puzzle, you have to complete the grid below so that every row, every column and every 3 x 3 block contains one each of the nine digits 1 to 9.

Always remember that this is a completely logical puzzle and you never have to guess. Only write a number in a square when you are certain by logical thinking that it is the only number that will go there.

The working boxes are there to help you to organise your thoughts. You only need to use them to write down the missing numbers if you are stuck.

	6			8		7		
2	5	7				4		9
9						1	3	
			2					3
8		5		6			2	
			3					8
		3		4				
1	2	4	8		7			
	8				6		7	

There are 52 numbers to find.

Tarquin Sudoku

To solve this puzzle, you have to complete the grid below so that every row, every column and every 3 x 3 block contains one each of the nine digits 1 to 9.

Always remember that this is a completely logical puzzle and you never have to guess. Only write a number in a square when you are certain by logical thinking that it is the only number that will go there.

The working boxes are there to help you to organise your thoughts. You only need to use them to write down the missing numbers if you are stuck.

3	6	8				5		1
1						2	4	
	7			9		8		
		4		5				
2	3	5	9		8			
	9				7		8	
			3					4
9		6		7			3	
			4					9

There are 52 numbers to find.

Tarquin
Sudoku

To solve this puzzle, you have to complete the grid below so that every row, every column and every 3 x 3 block contains one each of the nine digits 1 to 9.

Always remember that this is a completely logical puzzle and you never have to guess. Only write a number in a square when you are certain by logical thinking that it is the only number that will go there.

The working boxes are there to help you to organise your thoughts. You only need to use them to write down the missing numbers if you are stuck.

6		2	4	7	9			
3	5		2					
9				8			1	
					5		6	
			3	4	6	1		9
	9			1				8
		5				4		
	4		1		7		8	
		1				5		

There are 52 numbers to find.

Tarquin Sudoku

To solve this puzzle, you have to complete the grid below so that every row, every column and every 3 x 3 block contains one each of the nine digits 1 to 9.

Always remember that this is a completely logical puzzle and you never have to guess. Only write a number in a square when you are certain by logical thinking that it is the only number that will go there.

The working boxes are there to help you to organise your thoughts. You only need to use them to write down the missing numbers if you are stuck.

					6		7	
		4	5	7	2			1
	1			2				9
		6				5		
	5		2		8		9	
		2				6		
7		3	5	8	1			
4	6		3					
1				9			2	

There are 52 numbers to find.

Tarquin Sudoku

To solve this puzzle, you have to complete the grid below so that every row, every column and every 3 x 3 block contains one each of the nine digits 1 to 9.

Always remember that this is a completely logical puzzle and you never have to guess. Only write a number in a square when you are certain by logical thinking that it is the only number that will go there.

The working boxes are there to help you to organise your thoughts. You only need to use them to write down the missing numbers if you are stuck.

		1	3		6	8		
		3	2		5	7		
8								7
	7	6		5		2	8	
3		2				6		4
6		8				9		5
	5	7		1		3	6	
2								1

There are 51 numbers to find.

14

Tarquin
Sudoku

To solve this puzzle, you have to complete the grid below so that every row, every column and every 3 x 3 block contains one each of the nine digits 1 to 9.

Always remember that this is a completely logical puzzle and you never have to guess. Only write a number in a square when you are certain by logical thinking that it is the only number that will go there.

The working boxes are there to help you to organise your thoughts. You only need to use them to write down the missing numbers if you are stuck.

	4	7		9				2
	3	6		8				4
					8		9	
6			9	3		8		7
				7	5		4	3
				1	6		7	9
2			7	4		6		8
					2		3	

There are 51 numbers to find.

Tarquin Sudoku

To solve this puzzle, you have to complete the grid below so that every row, every column and every 3 x 3 block contains one each of the nine digits 1 to 9.

Always remember that this is a completely logical puzzle and you never have to guess. Only write a number in a square when you are certain by logical thinking that it is the only number that will go there.

The working boxes are there to help you to organise your thoughts. You only need to use them to write down the missing numbers if you are stuck.

				2	7		8	1
					3		4	
3			8	5		7		9
	4	7		9				5
	5	8		1				3
7			1	4		9		8
					9		1	
			8	6			5	4

There are 51 numbers to find.

Tarquin
Sudoku

To solve this puzzle, you have to complete the grid below so that every row, every column and every 3 x 3 block contains one each of the nine digits 1 to 9.

Always remember that this is a completely logical puzzle and you never have to guess. Only write a number in a square when you are certain by logical thinking that it is the only number that will go there.

The working boxes are there to help you to organise your thoughts. You only need to use them to write down the missing numbers if you are stuck.

	9	2					3	8
	5							4
8		1	4			9	6	
1		9	8			2	5	
	2							1
	6	5					9	7
		6		5	8		1	
		4		6	9		2	

There are 51 numbers to find.

17

Tarquin Sudoku

To solve this puzzle, you have to complete the grid below so that every row, every column and every 3 x 3 block contains one each of the nine digits 1 to 9.

Always remember that this is a completely logical puzzle and you never have to guess. Only write a number in a square when you are certain by logical thinking that it is the only number that will go there.

The working boxes are there to help you to organise your thoughts. You only need to use them to write down the missing numbers if you are stuck.

9	2						3	5
				9	3	6	7	
	1						9	4
			8		5		6	9
8						3		
	9	4	3		6			
	3				2			
5		2	6	4				
		6		3		9		2

There are 50 numbers to find.

Tarquin Sudoku

To solve this puzzle, you have to complete the grid below so that every row, every column and every 3 x 3 block contains one each of the nine digits 1 to 9.

Always remember that this is a completely logical puzzle and you never have to guess. Only write a number in a square when you are certain by logical thinking that it is the only number that will go there.

The working boxes are there to help you to organise your thoughts. You only need to use them to write down the missing numbers if you are stuck.

	4				3			
6		3	7	5				
		7		4		1		3
	2						1	5
1	3						4	6
			1	4	7	8		
		9		6			7	1
	1	5	4		7			
9					4			

There are 50 numbers to find.

Tarquin Sudoku

To solve this puzzle, you have to complete the grid below so that every row, every column and every 3 x 3 block contains one each of the nine digits 1 to 9.

Always remember that this is a completely logical puzzle and you never have to guess. Only write a number in a square when you are certain by logical thinking that it is the only number that will go there.

The working boxes are there to help you to organise your thoughts. You only need to use them to write down the missing numbers if you are stuck.

		5						4
			7	4	6	8		
4	2				8	5		
6		2	3					
7		5	4	2				
	8	9				2		5
2		8					1	7
		2		6			5	8
	5		1					

There are 50 numbers to find.

Tarquin Sudoku

To solve this puzzle, you have to complete the grid below so that every row, every column and every 3 x 3 block contains one each of the nine digits 1 to 9.

Always remember that this is a completely logical puzzle and you never have to guess. Only write a number in a square when you are certain by logical thinking that it is the only number that will go there.

The working boxes are there to help you to organise your thoughts. You only need to use them to write down the missing numbers if you are stuck.

				8	5	7	9	
5	3				9	6		
			6					5
3		9					2	8
	6			2				
			3		7		6	9
7		3	4					
	9	1				3		6
8		6	5	3				

There are 50 numbers to find.

Tarquin
Sudoku

To solve this puzzle, you have to complete the grid below so that every row, every column and every 3 x 3 block contains one each of the nine digits 1 to 9.

Always remember that this is a completely logical puzzle and you never have to guess. Only write a number in a square when you are certain by logical thinking that it is the only number that will go there.

The working boxes are there to help you to organise your thoughts. You only need to use them to write down the missing numbers if you are stuck.

	8			1	9	2		7
					4		3	
		1					8	4
	1			4		7		8
			1					
5	6				2	1		
1		3					4	
		9		6				
7	2			3	5	8		

There are 52 numbers to find.

Tarquin Sudoku

To solve this puzzle, you have to complete the grid below so that every row, every column and every 3 x 3 block contains one each of the nine digits 1 to 9.

Always remember that this is a completely logical puzzle and you never have to guess. Only write a number in a square when you are certain by logical thinking that it is the only number that will go there.

The working boxes are there to help you to organise your thoughts. You only need to use them to write down the missing numbers if you are stuck.

	2	1		9		3	8	
		5						4
			2				5	9
	5			2		8	9	
2								
		3		7	6	2		
			4		2			5
	7		1					
	4	6		3	8	9		

There are 52 numbers to find.

Tarquin Sudoku

To solve this puzzle, you have to complete the grid below so that every row, every column and every 3 x 3 block contains one each of the nine digits 1 to 9.

Always remember that this is a completely logical puzzle and you never have to guess. Only write a number in a square when you are certain by logical thinking that it is the only number that will go there.

The working boxes are there to help you to organise your thoughts. You only need to use them to write down the missing numbers if you are stuck.

3								
	6			3		9	1	
		4		8	7	3		
			5		3			6
	8		2					
	5	7		4	9	1		
	3	2		1		4	9	
			3				6	1
		6						5

There are 52 numbers to find.

Tarquin Sudoku

To solve this puzzle, you have to complete the grid below so that every row, every column and every 3 x 3 block contains one each of the nine digits 1 to 9.

Always remember that this is a completely logical puzzle and you never have to guess. Only write a number in a square when you are certain by logical thinking that it is the only number that will go there.

The working boxes are there to help you to organise your thoughts. You only need to use them to write down the missing numbers if you are stuck.

						4		
1	2		4					7
4			9	8			5	
		7		4	6			
					3			9
2			5	1			8	6
5	1		2				3	4
	7	2			4			
		6					7	

There are 52 numbers to find.

Tarquin
Sudoku

To solve this puzzle, you have to complete the grid below so that every row, every column and every 3 x 3 block contains one each of the nine digits 1 to 9.

Always remember that this is a completely logical puzzle and you never have to guess. Only write a number in a square when you are certain by logical thinking that it is the only number that will go there.

The working boxes are there to help you to organise your thoughts. You only need to use them to write down the missing numbers if you are stuck.

	7		3			6		9
6					7	8		1
		5		4				
	8		1			2	6	4
3								
	6		4			7	1	3
		4		7			3	
9					4	5		8
	3		6			4		2

There are 49 numbers to find.

Tarquin Sudoku

To solve this puzzle, you have to complete the grid below so that every row, every column and every 3 x 3 block contains one each of the nine digits 1 to 9.

Always remember that this is a completely logical puzzle and you never have to guess. Only write a number in a square when you are certain by logical thinking that it is the only number that will go there.

The working boxes are there to help you to organise your thoughts. You only need to use them to write down the missing numbers if you are stuck.

	8		1		7		4	
		7	2		9			8
6						5		
	9		5	7	3		2	
		4						
	7		4	2	8		5	
5				4		8		
		1	9		6			5
	4		3		5		7	

There are 49 numbers to find.

Tarquin Sudoku

F3

To solve this puzzle, you have to complete the grid below so that every row, every column and every 3 x 3 block contains one each of the nine digits 1 to 9.

Always remember that this is a completely logical puzzle and you never have to guess. Only write a number in a square when you are certain by logical thinking that it is the only number that will go there.

The working boxes are there to help you to organise your thoughts. You only need to use them to write down the missing numbers if you are stuck.

6				5		9		
		2	1		7			6
	5		4		6		8	
		8	3		1			9
	9		2		8		5	
7						6		
		5						
	8		5	3	9		6	
	1		6	8	4		3	

There are 49 numbers to find.

Tarquin Sudoku

To solve this puzzle, you have to complete the grid below so that every row, every column and every 3 x 3 block contains one each of the nine digits 1 to 9.

Always remember that this is a completely logical puzzle and you never have to guess. Only write a number in a square when you are certain by logical thinking that it is the only number that will go there.

The working boxes are there to help you to organise your thoughts. You only need to use them to write down the missing numbers if you are stuck.

6			7				1	
	2	8		3				7
	5	7			6	9		
	4	2		9				1
	3	9			1	6		
			8				7	
				6				
4	6	1			9	7		
9	7	5			2	4		

There are 49 numbers to find.

Tarquin
Sudoku

To solve this puzzle, you have to complete the grid below so that every row, every column and every 3 x 3 block contains one each of the nine digits 1 to 9.

Always remember that this is a completely logical puzzle and you never have to guess. Only write a number in a square when you are certain by logical thinking that it is the only number that will go there.

The working boxes are there to help you to organise your thoughts. You only need to use them to write down the missing numbers if you are stuck.

6								1
		8		4		3	6	
5				1		8		9
	6	3		9		7		5
2		1		5				
4								8
	9						8	
3		2	8			5		
		5			9	1		4

There are 51 numbers to find.

Tarquin Sudoku

To solve this puzzle, you have to complete the grid below so that every row, every column and every 3 x 3 block contains one each of the nine digits 1 to 9.

Always remember that this is a completely logical puzzle and you never have to guess. Only write a number in a square when you are certain by logical thinking that it is the only number that will go there.

The working boxes are there to help you to organise your thoughts. You only need to use them to write down the missing numbers if you are stuck.

				7		2		
		5			9		4	7
		2		6		1	9	
		1	7		4	6	8	
		6		3	2			
				5		9		
			1					9
9				4	3		6	
	1				6	5	2	

There are 51 numbers to find.

Tarquin Sudoku

To solve this puzzle, you have to complete the grid below so that every row, every column and every 3 x 3 block contains one each of the nine digits 1 to 9.

Always remember that this is a completely logical puzzle and you never have to guess. Only write a number in a square when you are certain by logical thinking that it is the only number that will go there.

The working boxes are there to help you to organise your thoughts. You only need to use them to write down the missing numbers if you are stuck.

				6		1		
	2	8		5	7	9		
	7		4	3				
1			5	4		7		
	2			7	6	3		
		2						1
			8		3			
	3		7		2	1		
	6			1		5	8	

There are 51 numbers to find.

Tarquin
Sudoku

To solve this puzzle, you have to complete the grid below so that every row, every column and every 3 x 3 block contains one each of the nine digits 1 to 9.

Always remember that this is a completely logical puzzle and you never have to guess. Only write a number in a square when you are certain by logical thinking that it is the only number that will go there.

The working boxes are there to help you to organise your thoughts. You only need to use them to write down the missing numbers if you are stuck.

7								2
	6	9			3		1	8
5	4				8			
6	5			2			8	
	8		3				4	7
		3				2		
9								4
8					4		2	3
	2				7	9	6	

There are 51 numbers to find.

Tarquin Sudoku

To solve this puzzle, you have to complete the grid below so that every row, every column and every 3 x 3 block contains one each of the nine digits 1 to 9.

Always remember that this is a completely logical puzzle and you never have to guess. Only write a number in a square when you are certain by logical thinking that it is the only number that will go there.

The working boxes are there to help you to organise your thoughts. You only need to use them to write down the missing numbers if you are stuck.

1								
	2	6	7				1	3
4	7				2	8	6	
5		3	1		7			
6					9		2	7
			6					
8			9		1	7		5
					8			
7	1		4		5			8

There are 50 numbers to find.

Tarquin Sudoku

To solve this puzzle, you have to complete the grid below so that every row, every column and every 3 x 3 block contains one each of the nine digits 1 to 9.

Always remember that this is a completely logical puzzle and you never have to guess. Only write a number in a square when you are certain by logical thinking that it is the only number that will go there.

The working boxes are there to help you to organise your thoughts. You only need to use them to write down the missing numbers if you are stuck.

				2				
	8		3		7	2	4	
		3	8	5		7		9
	2	8		6	4			
		1		7		3	8	
	7							
	1	2		9			6	8
		9						
	5	6	2	8			9	

There are 50 numbers to find.

Tarquin Sudoku

To solve this puzzle, you have to complete the grid below so that every row, every column and every 3 x 3 block contains one each of the nine digits 1 to 9.

Always remember that this is a completely logical puzzle and you never have to guess. Only write a number in a square when you are certain by logical thinking that it is the only number that will go there.

The working boxes are there to help you to organise your thoughts. You only need to use them to write down the missing numbers if you are stuck.

		1						
	2	3		1			7	9
	6	7	3	9			1	
	8							
	3	9		7	5			
		2		8		4	9	
		4	9	6		8		1
	9		4		8	3	5	
				3				

There are 50 numbers to find.